The Big Ditch Waterways

The Story of Canals

by Solveig Paulson Russell

illustrated by Don Sibley

A Finding-Out Book   *Parents' Magazine Press* • New York

Library of Congress Cataloging in Publication Data

Russell, Solveig Paulson.
   The big ditch waterways.

   (A Finding-out book)
   Includes index.
   SUMMARY: Discusses man-made canals used throughout the world for irrigation, transportation, and drainage.
   1. Canals—Juvenile literature. [1. Canals]
I. Sibley, Don.   II. Title.
TC745.R87   627'.13   76-46407
ISBN 0-8193-0898-6
10 9 8 7 6 5 4 3 2 1

Text copyright © 1977 by Solveig Paulson Russell
Illustations copyright © 1977 by Don Sibley
All rights reserved
Printed in the United States of America

Contents

INTRODUCTION   5
EARLY RIVER AND CANAL TRAVEL   6
DRAINAGE AND IRRIGATION CANALS   9
BUILDING CANALS   12
   Locks   13
THE ERIE CANAL   18
   Building   19
   Life on the Erie   23
OTHER AMERICAN CANALS   27
CANALS IN OTHER COUNTRIES   33
FAMOUS CANALS   39
   The Suez   39
   The Panama   45
   The Saint Lawrence Seaway   54
   "Soo" Canals   59
THANKS TO CANALS   61
INDEX   63

Introduction

When rain falls on the ground, it often makes mud puddles. Usually there are a number of puddles close together. If you take a pointed stick and draw it between two puddles, you will make a little ditch. Then the water can run from one puddle to the other. The ditch is a tiny canal. It joins the two puddles.

If the puddles are big, and if the ditch is made wide and deep enough, you could sail a leaf boat from one puddle to the other. The ditch canal would be like a water highway for the leaf.

Larger bodies of water, many millions of times bigger than rain puddles, are joined by canals. People have dug huge ditches between lakes, rivers, and even oceans. These canals are waterways. Large ships pass over them. The ships carry people and goods from one place to another.

Early River and Canal Travel

Before there were trains in our country, people had to travel on foot, by horseback, or with horses or mules and wagons. The towns and farms were far apart. The roads and trails were bumpy and rough. It was hard to get to a distant place, and the journey was often long.

People who lived near rivers found that they could use the rivers for traveling by boat or raft.

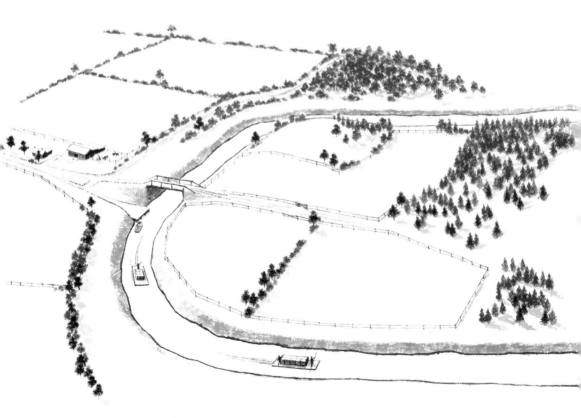

SIDE CANAL

They could get the things they had to sell to market by floating them down the rivers. This was much easier and faster than hauling them over poor roads.

But sometimes there were falls or rapids in the river, or big rocks blocked the way. Then the people dug big ditches, or canals, around the trouble spots. They made the water above a falls flow into the canal they dug at the side of the river. It ran through the canal and then ran back

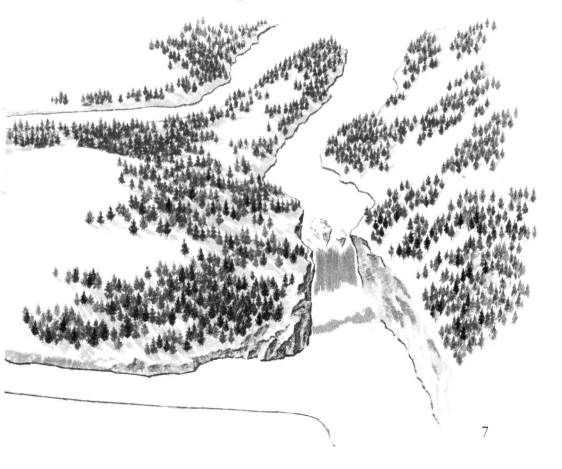

into the river below the falls. With these side canals, men could float their goods for many miles.

The canals around trouble spots in rivers were not very long. But later people saw that canals could be used to join big rivers or lakes. When they were built, these canals became many miles long. As more and more people moved westward, these longer canals were very important for moving the early settlers and their goods.

As far back as 2,000 years before Christ, Egyptians dug canals and used them for transportation. Since then, great numbers of canals, large and small, have been built in many countries.

Drainage and Irrigation Canals

Canals are sometimes used to drain away water from lands that are too wet to grow crops. When canals are dug in wet lands, the water goes into the big ditches, which are made lower than the top of the ground. Drainage canals make it possible to use land that would otherwise be wasted.

In some cities, canals carry off sewage after it has been treated so that it will no longer be a danger to health.

DRAINAGE CANAL

Canals are also important for bringing water for thirsty plants to dry lands. Bringing water to plants that need it is called *irrigation*. A great variety of plants now grow on land that once was too dry for them to live on.

Irrigation canals are usually dug from a man-made lake, or reservoir, which is a big hole that has been dug in the ground. The water that fills reservoirs comes to them from rain and snow high in the mountains. As the rain falls, or the

IRRIGATION CANALS

snow melts, it runs or trickles down into the reservoir. There it is held until the plants need water.

When the land dries in the hot sun, the water from the reservoir is let out into a canal. The water may travel long distances until it branches off into smaller canals. These smaller canals lead to the farmers' fields where even smaller ditches carry the water to the thirsty plants. Most irrigation canals in our country are in the West, where there is much dry land.

Building Canals

The earliest canals were dug by men using picks and shovels. It is not easy for us to understand how hard this work must have been, and how long it took. Today our machines—diggers and earth movers—do work in a short time that took thousands of hours in the past.

The width and depth of a canal depend on the boats that are expected to use it. The canals must be deep enough so that the water under the boats will float them. They must be wide enough so that ships may pass each other, going in opposite directions.

The sides of a canal may be sloping, but usually they are straight up and down. The earliest canals were made of rocks, sod, clay, or wooden planks. Now they are made of concrete.

LOCKS

The very first canals were made only on level ground. But with later canals, one body of water, such as a lake, might be much higher than another joined to it by a canal. In the 14th or 15th century, a way was found to float boats from one level of water to another, higher or lower. Both Italy and Holland claim it was thought of in their country, but there is no sure record of this. To move boats up or down a canal, a group of water stairways is used. The stairways are called *locks*. Locks make it possible for boats to travel on canals through high hills.

Canal locks are big box-like rooms, or chambers, with high walls, built as part of the canal. There are heavy water-tight gates at each end of the chambers. When a boat is traveling to a higher level, it first enters the lowest chamber. After it is in this chamber, the gate behind it is closed. Then more water is put into the chamber by valves or pumps. The water flows in, and as it becomes deeper it lifts the boat in the big room. This is the same way that a toy boat rises in a bathtub as the water flows in. When the first chamber is as full of water as the one above it, the gate in front of the boat is opened and the boat moves into the water of the next chamber.

In the second lock the same thing happens. Rising water lifts the boat upward. Thus it is raised so that it can travel on at levels much higher than where it first entered the locks.

When a boat is going from a higher to lower levels, the water in the chambers is slowly let out. Then the boat is lowered until it can pass into the next lower chamber and into the level part of the canal.

When two pairs of locks are built side by side, some boats can go in one direction at the same time that other boats go in the opposite direction.

If a canal must cross a valley, or gully, big bridges, called aqueducts, are used. For these, a bridging arch is built over the valley. The canal is made over the top of the arch. It flows along there much as a railroad's train tracks cross a trestle.

In canals, boats are usually not allowed to use their own power, or go fast. This is because their engines stir up the water so much that the sides and bottom of the canal would soon be worn away or damaged. So ships going through canals are usually pulled, or towed by engines on the banks. Mules and horses, and sometimes even men, used to walk along the canal banks pulling the boats with long tow ropes.

The Erie Canal

Canals helped the United States grow from a small country to a big one. As we have learned, the settlers used them, and rivers, for transportation when they could. As the young country grew, people wanted canals dug between rivers so that there would be longer waterways.

Some of the leaders of the country thought the cost of building long canals was too much. But one man who didn't think so was the governor of New York, DeWitt Clinton. Clinton saw that having more canals for transportation would be of great importance to many people. He became very much interested in them and worked hard to make one, especially, become a reality. He got the elected officials of New York State to agree to build a canal connecting the Great Lakes to the

Hudson River and the Atlantic Ocean. This was the Erie Canal, which took eight years to build. It was finished in 1825.

BUILDING

The men who built the Erie Canal went through many hardships. They had to work in deep, swampy ground where mosquitos were thick. They had to cut paths through heavy forests of big tall trees. They worked in heat and

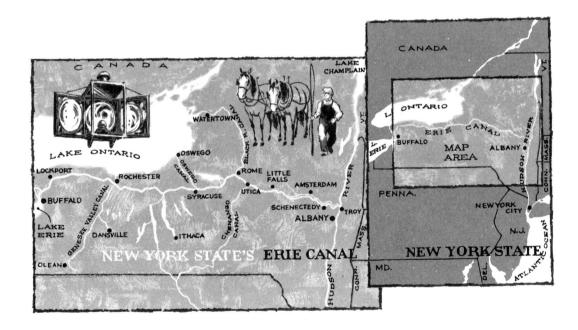

in freezing weather, and they had only hand tools and horse-drawn scrapers and plows to move dirt, rocks, and trees. Many times they were discouraged. Some people called the canal "Clinton's Folly," or "The Governor's Ditch." But at last the big ditch was finished. It was then 4 feet (1.2 meters) deep, 40 feet (12.2 meters) wide, and 363 miles (584.5 kilometers) long.

People gathered all along the banks for a big celebration. There were fireworks, speeches, bands, guns fired, and much shouting. Gaily decorated barges passed along the canal. Governor Clinton rode the whole length of the canal on one. When his barge reached the end of the canal, he emptied into New York harbor a keg of water taken from Lake Erie. This act was a symbol of the joining of

the Atlantic Ocean with the Great Lakes. The happy people called it the "Wedding of the Waters."

Soon the canal was a very busy waterway. And the money that had been spent to build it—more than seven million dollars—was quickly repaid by those who used it.

Horses walked along the towpaths beside the canal. They pulled passenger boats called packets. Mules pulled barges loaded with all kinds of things—chickens, stoves, clothing, and all the belongings of people going West to find new homes, or for those who had already settled.

Boys often drove the horses that pulled the boats. The boys were called "hoggees." Where this name came from is not known. But it has been suggested that it may have come from the fact that the boys often called to their animals to "Haw" or "Gee" when they wanted them to go in one direction or another, and the two words sounded something like "hoggee."

LIFE ON THE ERIE

New towns sprang up along the way although they were, at first, only clearings with a building or two where travelers could get supplies. Many bridges were built over the canal so that people could pass from one side to the other.

The slender passenger packets were the fastest boats on the canal. They were looked at, enjoyed, and used with pride. They had cabins with narrow, shelf-like places, one above another, where the passengers could sleep, but not with much comfort. The meals were good, however, and plentiful.

During the day the passengers could sit on the top of the cabin and enjoy the sights. They could even jump off the boat onto the towpath for a walk, and then jump back on again later as the boat moved slowly forward.

Whenever the packet came to a bridge, someone would cry, "Low bridge! Everybody down!" Then the people on top of the cabin would have to duck down, or they would be knocked over. Sometimes the passengers who were walking would go ahead to a bridge. They would jump on board from the bridge as the boat passed under it—an exciting but dangerous thing to do.

Boats that traveled both day and night had big lanterns at the front that were called *nighthawkers*.

These burned oil and gave a clear light that did not attract insects. The lights shone over the canal far enough so that the drivers on the towpaths could see if anything was in the way of the boat they pulled.

When boats of any kind came to locks, they had to wait their turn to go through. At such times, the boatmen and passengers visited with each other, did chores needed to keep things in order, and enjoyed the countryside.

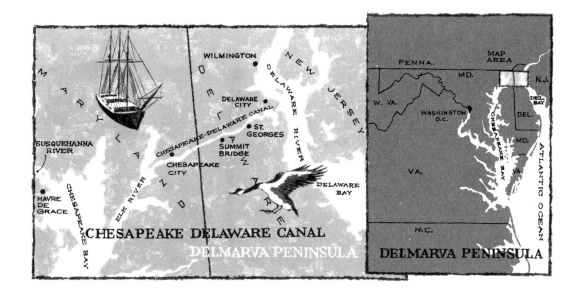

Other American Canals

With the success of "Clinton's Big Ditch," other American canals were built. Some of them were the Chesapeake and Delaware Canal, the Delaware and Raritan Canal, the Illinois and Michigan Canal, and the Chesapeake and Ohio Canal.

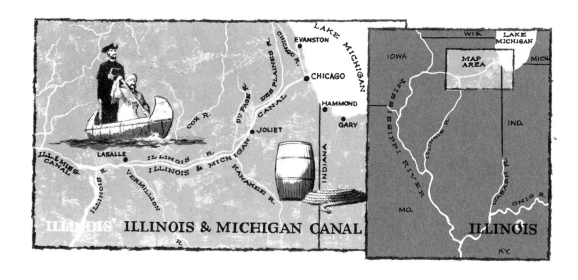

Another one, the Pennsylvania Canal, didn't use locks to get boats over the hills. The Allegheny Mountains it crossed were too high. The boats using this canal were built so that they could be unbolted into sections. Railroad tracks ran down into the canal on each side of the mountain range. Carriages, or cradles on wheels, ran down the track into the canal where a section of the boat would be floated onto the cradle. Then a

steam-powered engine hauled the loaded carriage up a slope to a shed.

When several loaded carriages were at the shed, a locomotive was hooked on to a string of them. The engine pulled the boat sections up over the mountains and let them down into the canal on the other side. There they were put together again to continue on their way by water.

The Pennsylvania Canal ran from Columbia, on the Susquehanna River, to Pittsburgh. The mountain distance was about 36½ miles (58.8 kilometers) long. It was opened in 1834.

When the canals were in greatest favor—in the early 1850s—there were about 4,500 miles (7,248 kilometers) of canals in use, but thirty years later half of them were no longer used.

Many of the once busy canals all over the country were left to dry up, fall in, or fill with weeds. Some that had been started were left unfinished. Without care, locks and dams crumbled

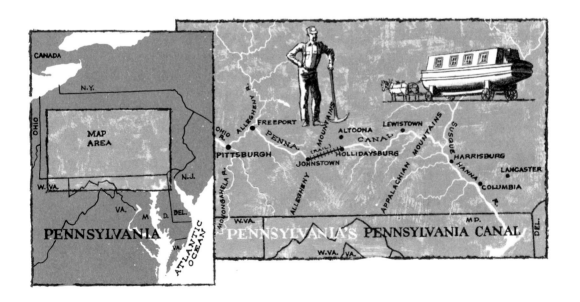

away, and abandoned barges were left to rot.

The coming of the railroads was the reason for the abandoning of many canals. Railroads could go over hills and mountains on tracks. They could run across desert lands where there was little or no water. They were faster than canal travel—much faster. The railroads carried people and goods into new areas where canals could never go. Some canals froze over in winter so that they couldn't be used. Trains could run in any weather.

But later, in the early 1900s, large canals once again became important for moving all kinds of heavy goods. That was because it costs less to ship heavy things like ore, coal, and material needed for construction jobs—such as road building—by water than by train or truck.

So many of the canals still in use were widened and made deeper. Old locks were rebuilt. The canals were made so that big barges could move by their own engines or by tug boats. Some parts of the old Erie Canal became a part of a much longer canal—the New York State Barge Canal. This waterway is over 500 miles (805 kilometers) long, and can handle large ships.

Other canals and rivers, too, were improved so that bigger boats could travel on them. New canals were dug that allowed ships to go from the oceans to inland cities. Now such cities as Houston, Texas, and Seattle, Washington, have direct connections with ocean ship travel.

Canals in Other Countries

In Europe and other places of the world, canals served people well, and have done so for centuries.

The Grand Canal of China is a very old one. It is believed to have been started about seven hundred years before Christ. It was finished about two thousand years later.

There are nearly 200,000 miles (322,000 kilometers) of canals in China. Many families spend their lives on canal boats there. The children help on the boats, doing the ordinary chores that most children do on land.

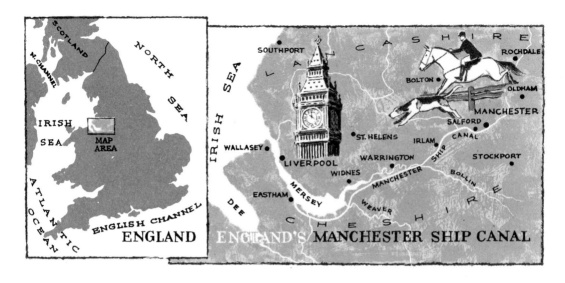

The British, too, have many canals. One of England's most important ones is the Manchester Ship Canal which connects the sea with the city of Manchester 35 miles (56 kilometers) inland. This canal is an example of how a canal can make an inland city become an important seaport.

The Grand Canal between Dublin and Ballinasloe, in Ireland, is 165 miles (266

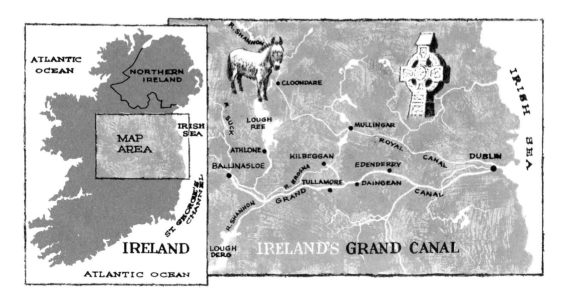

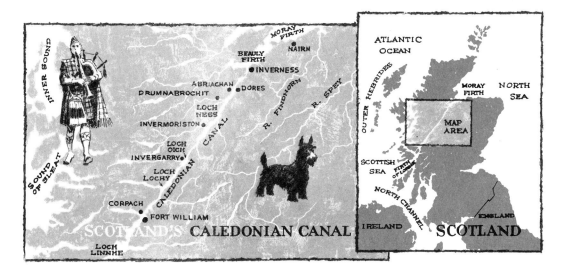

kilometers) long. In Scotland, the Caledonian Canal extends across the land from Inverness to Fort William.

The big land of the Soviet Union has canals that unite its widely separated parts. The Volga-Baltic, opened in 1964, is one of these. Another is the White Sea-Baltic Canal between Leningrad and Archangel.

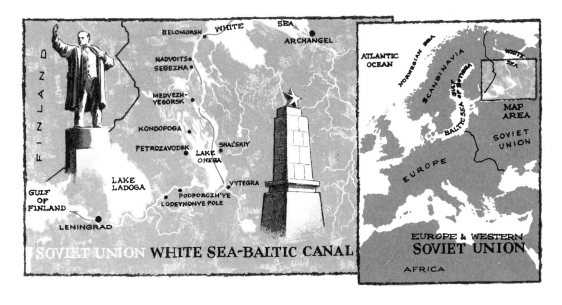

The Netherlands is a network of canals. There canals are often used as we would use ordinary streets. Both the Netherlands and Belgium are low countries. The sea would wash much of the land away if it were not for the dams, or dikes, that are built. So canals in these places serve not only for transportation, but they also help to drain the country.

Many people think of the city of Venice in Italy when they think of canals. There the streets are canals with bridges crossing from one bank to another. Colorful pointed boats, called gondolas, are used as taxis and for pleasure there.

France has more than 3,000 miles (4,830 kilometers) of waterways. Many were built long

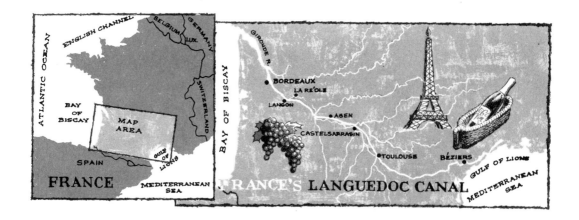

ago. One of these, the Languedoc Canal, was completed in 1668. It connects the Mediterranean Sea with the Bay of Biscay.

The Nord-Ost-See-Kanal is another name for the German Kiel Canal. It connects the Baltic and North Seas and is Germany's most important canal. Another German canal is the Hohenzollern Canal, which joins the Oder and Spree rivers and makes a ship channel between Berlin and Stettin on the Baltic Sea.

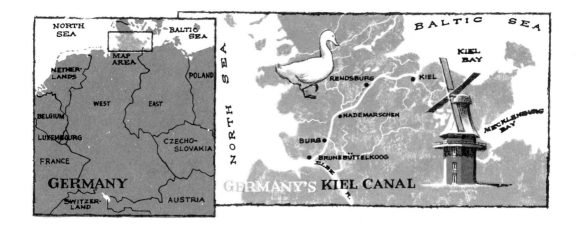

Famous Canals

SUEZ

All canals are important to those who live near them. But some are important to people all over the world, such as the Suez and Panama canals.

The Suez Canal was the first big ship canal made in recent times. It was opened in 1869 and it runs from the Mediterranean Sea to the Red Sea through Egypt. It crosses the Isthmus of Suez, a narrow piece of land between the two seas. It is one of the busiest canals made between oceans.

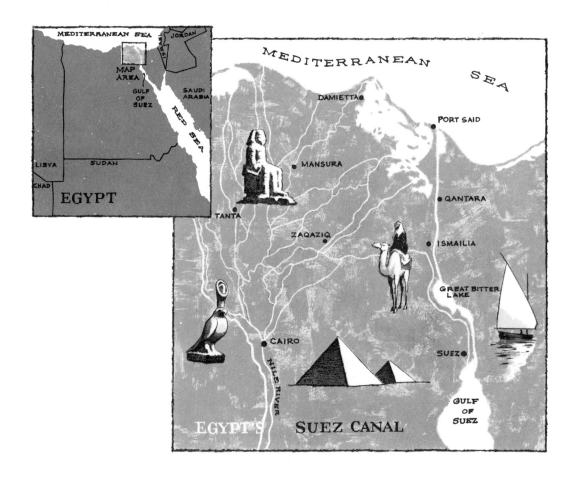

The Suez Canal has no locks. It is built at sea level. It is 103 miles (166 kilometers) long, 36 feet (10.9 meters) deep, and 196 feet (60 meters) wide.

For centuries people had known that if such a canal could be made it would be a wonderful shortcut. It would mean that the ships going to the Orient from Europe and other places wouldn't have to make the long trip around the southern tip of Africa.

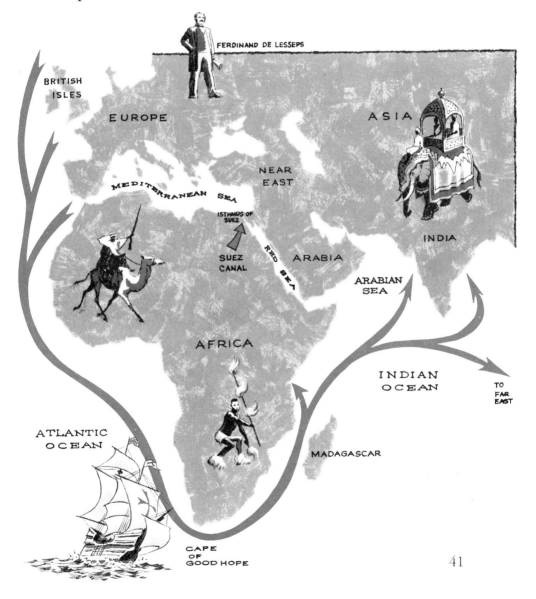

Ferdinand de Lesseps, a French diplomat and engineer, thought about, dreamt about, and studied how to build a canal across the isthmus. He worked tirelessly to get enough money to build the canal, and he stirred up interest in the project everywhere.

After de Lesseps had the money he needed, and the route the canal was to follow was decided upon, the work began in April, 1859. Ten years later the canal was open to travel.

The ten years of work were filled with many hardships. The actual digging was done in the merciless heat of the desert. At first, sand, rock, and swamp mud were moved slowly by dredges, and men and camels carried the stuff in baskets. Later steam shovels were used.

During the building, wars interrupted operations and caused greater hardships. Sand constantly blew in from the plains of the desert on each side of the canal and its removal added to the cost. When the canal was opened, in 1869, it had cost more than twice as much as everyone had expected.

The opening was marked with great celebrations. A convoy, or group of 68 vessels, decorated with flags, sailed south from Port Saïd (see page 40), the first convoy to use the canal. The Empress Eugénie of France was on the leading yacht. Thousands of important people were in the procession.

When the ships were halfway through the canal, they stopped at Ismailia (see page 40) to attend celebrations. In the evening of the opening day, a grand ball was held with 6,000 guests present, including the Emperor of Austria, and other royal guests. All of them paid tribute to the work of Ferdinand de Lesseps, a man who had worked hard and long to make a dream come true.

Since the opening, wars and other troubles have affected the use of the Suez Canal, and the part of the world where it lies has often been disturbed by people fighting for power.

THE PANAMA CANAL

Joining North and South America, there is a narrow piece of land called the Isthmus of Panama. The Atlantic Ocean is on one side and the Pacific Ocean is on the other side. Before the Panama Canal was built through the isthmus, all boats going from one ocean to the other had to go around Cape Horn at the tip of South America (see page 46). This was a long way. So the idea of cutting through the narrow piece of land, the isthmus, had long been dreamed about, just as had been the case with the Suez Canal.

Vasco Balboa, a Spanish explorer who visited

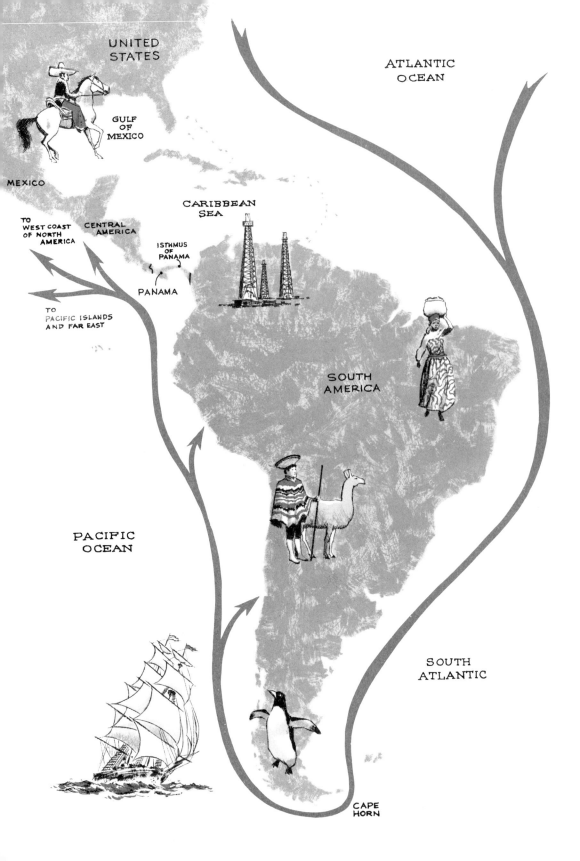

the isthmus as far back as 1517, thought it would be possible to join the two oceans by some kind of a waterway. Both the United States and Great Britain considered making a canal to do this in the early 1800s, but because of disagreements, the plan was not carried out.

Finally, in 1882, a French company began the digging of the canal. The company was headed by Ferdinand de Lesseps, the man who had made a name for himself with the Suez Canal. He was 74 years old. This company at first planned to dig the whole canal at sea level so that they would need no locks. But the plan didn't work. The Isthmus of Panama is not flat country as the Isthmus of Suez is. It has high hills that had to be dug through, or over which ships would have to be lifted.

The French planned carefully, but their engineers didn't have the tools they needed for such a big job. Money was wasted and some of it was taken by dishonest men connected with the work. Worst of all, the mosquitoes that infested the swampy land carried disease germs to the workers. Many of them died, and scientists did not know then how to fight these tiny insect killers.

De Lesseps' company went bankrupt in 1889. After this, until about 1904, the plan of building the canal met with many difficulties, but finally the United States gained the right, from the country of Panama, to build and govern the canal.

In 1904, a doctor, Colonel William C. Gorgas, and the United States Army Engineers began to clean up the mosquito-filled swamps. Gorgas had already gained fame through his work in wiping out yellow fever in Cuba. For two years, most of the work on the canal was the draining of land, cutting of brush, and destroying large areas of grass where the mosquitoes swarmed, and also cleaning up living conditions. With this work, Colonel Gorgas and his men wiped out yellow fever mosquitoes and the rats that carried a plague, and subdued the malaria-carrying mosquitoes. The work on the canal could now go forward.

The Panama Canal has been called man's greatest engineering achievement. It is about 50 miles (81 kilometers) long. It goes from the Bay of Panama on the Pacific to Limón Bay on the Atlantic. It is called a "lock and lake" waterway.

This is because the builders made a big lake, Lake Gatun, about 85 feet (26 meters) above sea level and then made three locks on each side of the lake so ships can pass from one ocean to the other.

Colonel George W. Goethals of the Army Engineer Corps was in charge of the building. He was a forceful man of much ability. He was a good organizer and a leader of men.

One of the most troublesome parts of the building of the Panama Canal was through a

section where the land is soft and slippery. When workmen cut through it, the sides constantly kept sliding back into the places they dug. This place was first known as the Culebra Cut, but now its name has been changed to the Gaillard Cut (see page 49). David Gaillard was an engineer in charge of digging. The cut is 9 miles (14.5 kilometers) long.

Most of the work on the canal was finished in 1914. On August 15 of that year a passenger ship

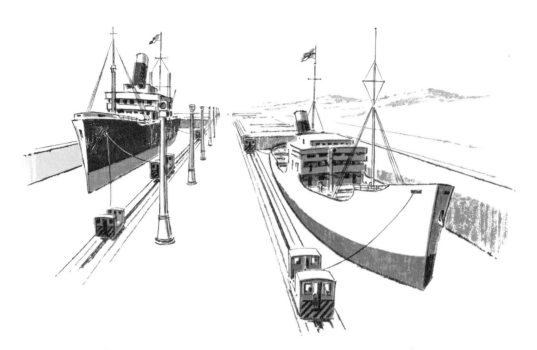

made the first complete journey through the canal. But the canal was closed for a number of months because of a huge landslide in the Gaillard Cut. On July 21, 1920 President Wilson proclaimed the Panama Canal open to all.

About 40,000 men from all over the world worked to cut through the jungles, hills, and swamps to make the Panama Canal. Their work has given the world an important waterway where millions of tons of cargo pass each year. Because of the Panama Canal, ships of all nations pass from one ocean to the other in a few hours, saving time, effort, and money.

Each of the ships that passes through the canal pays according to its size and the amount of its cargo. The United States maintains and governs the Panama Canal Zone, where the canal is built. The country of Panama gave the United States the right to govern it and to operate the canal by a treaty in 1903. The Canal Zone is governed through a company. This company takes care of harbors and a railroad that runs across the Isthmus of Panama. It takes care of telephones, electric and water systems, and the stores where people who live there can buy things they need. The company owns all the houses and rents them to the workers who keep the Panama Canal in

good shape. It pays all the salaries and bills to keep the canal always ready for use.

THE SAINT LAWRENCE SEAWAY

The building of the Saint Lawrence Seaway was a big job, second only to the building of the Panama Canal. The United States and Canada built it together. It is over 2,400 miles (3,865 kilometers) long, including the Great Lakes.

The Saint Lawrence River system was used for shipping by all but the largest ships before the Seaway project was begun. Both countries had built smaller canals through the years, but these were not large enough for the largest ships to travel through.

An International Joint Commission was organized, with three members from Canada and three from the United States, to develop better waterway travel on the Saint Lawrence. In 1921 the Commission recommended that the two countries together build a power-and-navigation project.

The U.S. Army Corps of Engineers directed the American work under a corporation, and the Saint Lawrence Seaway Authority of Canada directed the Canadian work. Both countries agreed on the fees ships would pay for use of the Seaway. These fees are used to pay for the building of the Seaway.

Now the completed Saint Lawrence Seaway joins the Great Lakes with the Atlantic Ocean and allows seagoing vessels to travel into the heartland of North America. The work on it

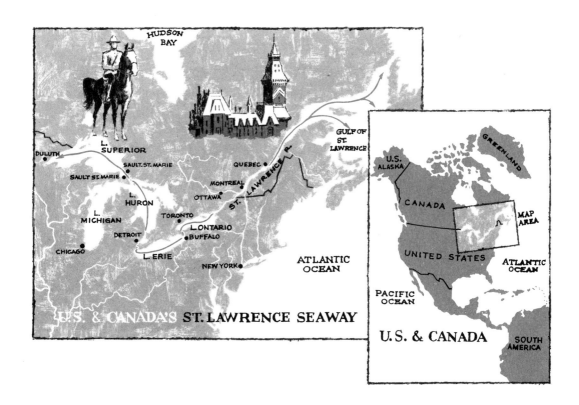

began in 1954. It was officially opened in June 1959. A huge hydroelectric plant built with the Seaway began serving electricity to both countries in 1958. Electric power from this big electric plant goes to New York, Vermont, and Ontario.

As with any such work, there were problems to be met and solved. It was found, for example, that some of the earth and rock to be removed was extremely hard—much harder than the builders had thought possible. Shovel teeth and bulldozer blades used on the hard material wore out in a few days, or hours, though they would have lasted for a year in ordinary soil. Drills wore out just as quickly. Of course this increased the cost of digging the tough canal sections.

The workers of one company used a jet-impelled boring machine. It sent a 4,000-degree flame into the rock. When the rock was white-hot, the men doused it with water. The water turned to steam and its force blew the rock apart.

Another problem was that the Seaway

construction had to take up home land where people had lived for many years. The land was needed for the building of the Seaway and for making reservoirs and lakes to control the flow of water, and for power dams. The people often did not want to give up their homes, even though they were paid well for them. The Seaway builders did all they could to help these people. They even moved old settlements and towns house by house and set them down in a new location.

The greatest number of giant machines ever to work together worked on the Saint Lawrence Seaway. One huge digger, called "The Gentleman," moved two million tons of earth in half a year.

During the winter, the cold weather slowed down the work. But still some workers kept on. They opened and shut locks to keep water moving. They used big heaters of different kinds, and even poured some concrete that they had to spray with steam to keep it from freezing.

When the Saint Lawrence Seaway was opened, people said it was like adding another coastline to North America. By using it, big ships can travel far inland to load grain and other goods from the inland areas and take them directly to many places in the world. Inland cities on the Seaway have become international ports.

The building of this great Seaway is an example of how nations working together can produce something that will benefit all who use it.

"Soo" Canals

At the far end of the Saint Lawrence Seaway lies Lake Superior. This lake is connected with Lake Huron by the famous Saulte Saint Marie locks that are commonly called the "Soo" locks These are two short canals running side by side. One of them is owned by the United States.

Work on them was finished in 1895. They are now a part of the Saint Lawrence Seaway.

More goods pass through the "Soo" locks than through any other locks in the world. Ore from Minnesota and Michigan mines goes through the locks to steel mills of Indiana, Ohio, and Pennsylvania. Coal and oil are carried westward. Wheat and other grain, lumber, oil, and many heavy machines go through these locks to big cities everywhere.

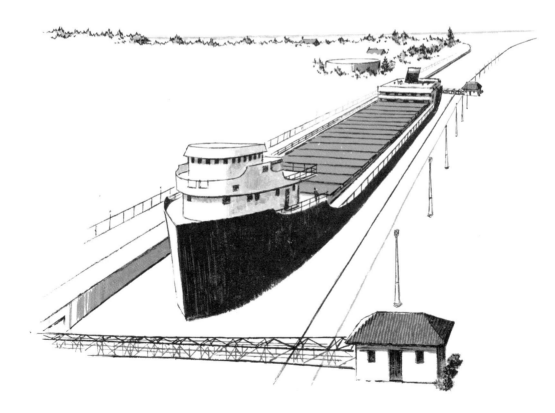

Thanks to Canals

In this book we have looked briefly at some of the world's many canals. We have learned something about the problems of building them. We have seen that these man-made waterways have been, and are, important to people's lives.

Irrigation canals help raise plants that could not be raised without the water they bring, thus making more food for more people. Drainage canals have made more land usable.

Canals have helped settlers in our country move into new areas that were hard to reach. They have made it possible for farmers and others using water transportation to get products to markets in distant places. And some canals have opened up new waterways so large ships can now journey far inland. Thus canals have created large cities, not located on the oceans, as shipping places where manufacturing of many products can also be carried on.

Shipping heavy goods by waterways is much cheaper than shipping by any other way, although it is slower, so canals save shipping costs.

Trains, trucks, and airplanes serve us well, and are much faster than canal travel, but the use of man-made waterways is of benefit to all.

Index

Africa, 41
Allegheny Mountains, 28
aqueducts, 16
Archangel, 35
Atlantic Ocean, 19, 22, 45, 49, 55
Austria, Emperor of, 44

Balboa, Vasco, 45, 47
Ballinasloe, 34
Baltic Sea, 38
barges, 21, 22, 31, 32
Bay of Biscay, 38
Bay of Panama, 49
Belgium, and canals, 36
Berlin, 38
boats, and canals, 5, 12, 13, 17;
    and locks, 13-15;
    towing of, 17;
    and Erie Canal, 21-22, 24-26;
    and Pennsylvania Canal, 28-29;
    to inland cities, 32, 55, 59, 61
bridges, 16, 23, 25

Caledonian Canal, 35
Canada, and St. Lawrence Seaway, 54, 55
Canal Zone, 53
canals, early, 7-8, 12;
    building of, 12-17, 19-20, 42, 46-52, 56-59;
    lack of use of, 30-1;
    rebuilding of, 32;
    see also different kinds of
Cape Horn, 45, 46
Chesapeake and Delaware Canal, 27
Chesapeake and Ohio Canal, 27
China, canals in, 33
clay, canals of, 12
Clinton, DeWitt, 18, 21
"Clinton's Big Ditch," 27
"Clinton's Folly," 20
Columbia, Pa., 30
concrete, canals of, 12
Cuba, 48
Culebra Cut, 51

dams, 30, 36
Delaware and Raritan Canal, 27
de Lesseps, Ferdinand, and Suez Canal, 42, 45;
    and Panama Canal, 47, 48
dikes, 36
drainage canals, 9, 36, 61
Dublin, 34

Egyptians, and canals, 8
electricity, and St. Lawrence Seaway, 56
England, canals in, 34;
    see also Great Britain
Erie Canal, 18-26, 32;
    building of, 19-20;
    opening of, 21-22;
    life on, 23-26
Eugénie, Empress, 43
Europe, 41;
    canals in, 33, 34, 36-38

fees, on canals, 22, 53, 55
Fort William, 35
France, canals in, 37-38;
    and Panama Canal, 47

Gaillard Cut, 49, 51
Gaillard, David, 51, 52
"Gentleman, The," 58
Germany, canals in, 38
Goethals, George W., 50
gondolas, 37
Gorgas, William C., 48
"Governor's Ditch, The," 20
Grand Canal of China, 33
Grand Canal, Ireland, 34
Great Britain, and Panama Canal, 47;
    see also England
Great Lakes, 18, 22, 55

hoggees, 22
Hohenzollern Canal, 38
Holland, *see* Netherlands
homes, moving of, 57
horseback, 6
horses, and wagons, 6;
    for towing, 17, 22
Houston, Texas, canals and, 32
Hudson River, 19

Illinois and Michigan Canal, 27
inland cities, canals and, 32, 34, 55, 59, 60, 61
International Joint Commission, 54
Inverness, 35
Ireland, canals in, 34
irrigation canals, 10-11, 61
Ismalia, 44
Italy, and locks, 13;
    canals in, 37

Kiel Canal, 38

63

Lake Erie, 21
Lake Gatun, 50
Lake Huron, 59
Lake Superior, 59
lakes, joining of, 8
Languedoc Canal, 38
leaf boat, sailing of, 5
Leningrad, 35
Limon Bay, 49
locks, 13-15, 26, 30, 32;
  Suez Canal and, 40;
  Panama Canal and, 47, 50;
  "Soo," 59-60

machines, for St. Lawrence Seaway, 56, 58-59
Manchester Ship Canal, 34
maps, 19, 27, 30, 34, 35, 38, 40, 41, 46, 49, 55
Mediterranean Sea, 38, 39
mosquitoes, and canal-building, 19, 48-49
mules, and wagons, 6;
  for towing, 17, 22

Netherlands, and locks, 13;
  canals in, 36
New York, electricity for, 56
New York harbor, 21
New York State Barge Canal, 32
nighthawkers, 25-26
Nort-Ost-See-Kanal, 38
North America, 45
North Sea, 38

oceans, joining of, 45-46, 52
Oder River, 38
Ontario, electricity for, 56
Orient, 41

Pacific Ocean, 45, 49
packets, passenger, 22, 24-25
Panama, and Panama Canal, 48, 53
Panama, Isthmus of, 45-47, 53
Panama Canal, 39, 45-54;
  building of, 46-52
Pennsylvania Canal, 28-30
Pittsburgh, Pa., 30
plants, irrigation of, 10-11
Port Saïd, 43
puddles, 5

raft, travel by, 6
railroad, 6, 16, 28, 31, 53

reservoirs, 10-11, 57
rivers, travel on, 6-7;
  joining of, 8, 18
rocks, canals of, 12

St. Lawrence River, 54
St. Lawrence Seaway, 54-60;
  building of, 56-59
St. Lawrence Seaway Authority of Canada, 55
Saulte Saint Marie, see "Soo" Canals
Scotland, canal in, 35
Seattle, Washington, canals and, 32
settlers, 8, 18, 61
sewage canals, 9
ships, see boats
side canals, 7-8
sod, canals of, 12
"Soo" canals, 59-60
South America, 45, 46
Soviet Union, canals in, 35
Spree River, 38
Stettin, 38
Suez, Isthmus of, 39, 42, 47
Suez Canal, 39-45, 47;
  and locks, 39-40;
  building of, 42;
  opening of, 43-45
Susquehanna River, 30

trains, see railroad
transportation, 5-8, 18, 22, 31, 32, 36, 60-62
travel, in early days, 6-7;
  ocean, 32

United States, and Panama Canal, 47, 48, 53;
  and St. Lawrence Seaway, 54, 55;
  and "Soo" locks, 59
U.S. Army Engineers, 48, 50, 55

Venice, canals in, 37
Vermont, electricity for, 56
Volga-Baltic Canal, 35

wagons, travel by, 6
"Wedding of the Waters," 22
White Sea-Baltic Canal, 35
Wilson, Woodrow, 52
wood, canals of, 12

yellow fever, 48